FIRST FACTS ABOUT

THE

SOLAR SYSTEM

Written by Michael Teitelbaum

Illustrations on pages 2–3 and 12–13 by Jon Friedman

Photographs courtesy of the National Aeronautics and Space Administration

Prepared with the cooperation of
James S. Sweitzer, Ph.D., Astronomer

Copyright © 1991 by Kidsbooks Inc.
7004 N. California Ave.
Chicago, IL 60645

Mercury
3,030 mi.

Venus
7,543 mi.

Earth
7,926 mi.

Mars
4,219 mi.

•**Jupiter**
88,600 mi.

You may think of the Earth as a very large place. To travel from one country to another can take many hours, even days. However, when we think of our planet's place in the universe, the Earth begins to look smaller and smaller. The universe is made up of billions of galaxies. Each galaxy is made up of hundreds of billions of stars. Many of these stars probably have planets spinning around them. The group of planets spinning around a star is called a Solar System. Our Sun is a star located in a galaxy called the Milky Way. It has nine planets spinning around it.

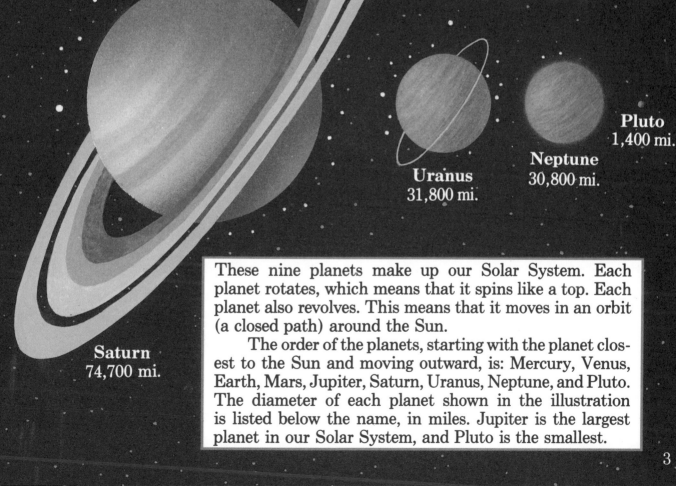

Saturn
74,700 mi.

Uranus
31,800 mi.

Neptune
30,800 mi.

Pluto
1,400 mi.

These nine planets make up our Solar System. Each planet rotates, which means that it spins like a top. Each planet also revolves. This means that it moves in an orbit (a closed path) around the Sun.

The order of the planets, starting with the planet closest to the Sun and moving outward, is: Mercury, Venus, Earth, Mars, Jupiter, Saturn, Uranus, Neptune, and Pluto. The diameter of each planet shown in the illustration is listed below the name, in miles. Jupiter is the largest planet in our Solar System, and Pluto is the smallest.

The Sun

The Sun is the center of our Solar System. All of the planets revolve around it. The Sun's diameter is 865,000 miles. One million Earths could fit inside the Sun! The Sun is a huge ball of burning gases, mostly hydrogen and helium. This burning ball gives off light and heat, lighting and heating all the planets in our Solar System. The Sun's gravity is what keeps the planets in their orbits. Without it the planets would just travel in straight lines out into space. The gravity constantly pulls the planets toward the Sun, always keeping them in their orbits.

The Sun is made up of layers. All we can see are the outer layers. The outermost layer, called the Chromosphere, contains the solar flares that are sometimes seen shooting out into space from the Sun's surface. The temperature of the Chromosphere is about 14,000°F. The layer just below the Chromosphere is the Photosphere. This gaseous layer is grainy and looks somewhat like an orange peel. The temperature in the Photosphere is 10,300°F. Certain patches on the Photosphere are cooler and darker than the rest of this layer. These dark patches are called sunspots. The larg-

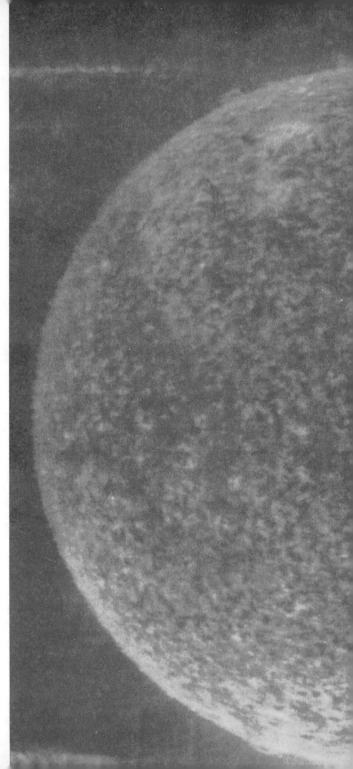

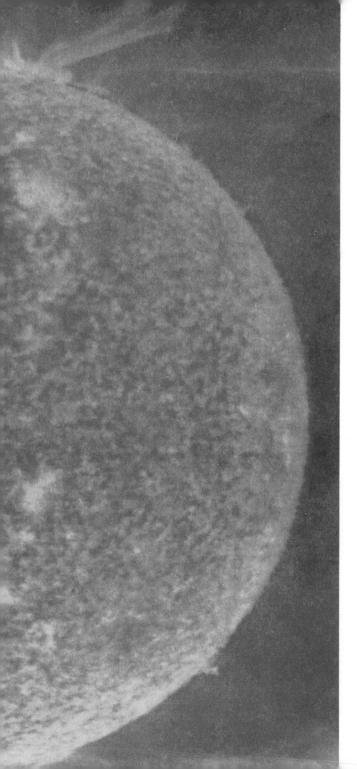

est sunspot ever seen was 150,000 miles long—20 times the diameter of the Earth!

Beneath these two visible outer layers of the sun are two layers made mostly of hydrogen gas. Finally we reach the Sun's core, which is made mostly of helium. The core is like an atomic furnace bursting with constant nuclear explosions. The core gives off the energy that creates the Sun's heat and light. The temperature at the core is about 27 million°F!

Like the planets, the Sun also spins (rotates). It takes 27 days for the Sun to complete one rotation.

The Sun was formed 5 billion years ago. It is believed that the Sun began as a huge cloud of whirling gas and dust in space. The cloud began to flatten out, and the gas and dust began to clump together into a giant fiery ball. Gravity squeezed the ball tighter and tighter, and it got hotter and hotter. Eventually the star was formed. Scientists estimate that our Sun will burn for another 5 billion years. Eventually it will run out of nuclear fuel. The outer layers will expand and drift off into space, leaving a tiny hot core that will cool down very slowly until it is invisible. Astronomers call this Earth-sized core a white dwarf star.

Mercury

Mercury is the planet closest to the Sun. It revolves faster than the other planets, taking only 88 days to complete its journey around the Sun. However, Mercury rotates so slowly that it takes 59 Earth days for Mercury to spin around completely one time!

The surface of Mercury is rocky and covered with craters. The largest crater is called Caloris Basin. This 800-mile-wide hole is bigger than the state of Texas! Mercury has no atmosphere, no water, and no moons. It is a lifeless rock like our own moon, with no atmosphere. Because of this, the sky on Mercury is always black.

Because it is so close to Earth, Mercury was one of the first planets discovered, thousands of years ago, in ancient times. But we really did not get a good look at it until 1974, when the Mariner 10 space probe took the first pictures of Mercury's surface.

The extremes of temperature on Mercury are very severe. On the side of the planet that faces the Sun, temperatures can soar to 870°F. On the side that faces away from the Sun, the temperatures can drop to 300°F below zero!

Venus

Venus, the second planet in the Solar System, is almost as large as Earth. It also comes closer to Earth than any other planet, and therefore is the brightest object in our night sky, except for the moon. Venus is, however, very different from Earth. The air on Venus is filled with carbon dioxide gas, which would suffocate humans. The surface is a rocky desert with high mountains. The only water on Venus is in the thick, steamy clouds that cover the planet. Once, Earth had similar clouds in its atmosphere, but as the temperature on Earth cooled down, the clouds turned into liquid water, forming the oceans, lakes, and rivers of our planet. Venus is too close to the Sun for this to happen there.

It takes Venus 225 Earth days to revolve around the Sun, but it takes 117 Earth days for Venus to rotate once. Venus rotates in the opposite direction from all the other planets. Because of this, the Sun rises in the west and sets in the east.

Like Mercury, Venus was discovered in ancient times. The planet was named after the Greek goddess of love. Venus has no moons. It is the hottest planet, with temperatures reaching 900°F. Even though Mercury is closer to the Sun, the steamy clouds on Venus act like a blanket, holding in the heat.

8

Earth

Earth is the only planet in our Solar System that we believe can support life. This is because the Earth is an ideal distance from the Sun. If it were any closer to the Sun, the temperatures would be too high to support life. It it were farther from the Sun, we would not have enough warmth and light for life here. Living things also need water and an atmosphere in which to breathe. Earth provides both of these things.

Earth's atmosphere is mostly made up of nitrogen and oxygen. Our atmosphere gives us the air we breath and protects us from harmful rays given off by the Sun. Billions of years ago, before the Earth cooled, the atmosphere was unbreathable, like the atmosphere on Venus. Then, as the Earth cooled, water in the clouds condensed into rain, forming oceans and other bodies of water. As the oceans formed, they absorbed the carbon dioxide, which humans can't breathe. Organisms in the ocean used the carbon dioxide to make their food, and gave off oxygen in the process. Eventually Earth's atmosphere became so filled with oxygen that it could support the life we know today.

The Earth itself is a giant ball of rock and metal. The outer layer is called the crust. The crust is a layer of rock 20 miles thick. On top of the crust lie the land and the oceans. Three-quarters of the Earth's surface is covered with water. The crust floats on an 1,800-mile-thick second layer of rock called the mantle. The mantle is under extreme heat and pressure. Even though it is made of rock, the heat and pressure make it soft, like clay. The mantle flows slowly beneath the harder surface of the crust. Under the mantle is the Earth's center, called the core. The core is a ball, composed of iron and nickel, and is as hot as the surface of the Sun.

A force called gravity, created by the mass of the planet, holds objects (and people) on the surface. That is why, when you throw something up in the air, it falls back to Earth, instead of just floating up into space.

The Earth has just one moon. It revolves around the Earth, just as the Earth revolves around the Sun. Although the Earth has many different climates, the temperature remains within a range of 135°F (in the deserts) to 126°F below zero (at the North and South Poles). It takes our planet 24 hours to complete one rotation (spin). This makes up one day. It takes 365¼days for the Earth to revolve around the Sun. This makes up one year.

Earth's Moon

The Moon is our closest neighbor in space. Its distance is only 239,000 miles from Earth—a short distance compared to the planets. The Moon was the first stop in man's exploration of other celestial bodies. During the 1960s and '70s the United States sent 9 manned missions to explore the Moon. We have learned a great deal from the astronauts who walked on the lunar surface.

The surface of the Moon is similar to those of Mercury and Mars—rocky and covered with craters, formed by meteors that crashed into its surface. The astronauts discovered that the Moon is very old. They found Moon rocks that date back 4.5 billion years!

It takes 27 days, 8 hours for the Moon to spin around once. This is also the time it takes for the Moon to revolve around the Earth. Because of this, only one side of the Moon ever faces the Earth. In addition to numerous craters, the side that faces us contains long, flat plains, covered with rocks and dust. Some sections of these plains appear darker than others. People used to believe that these dark areas were filled with water, so they were named maria, which means "seas." We now know that these dark areas are dry and dusty.

The far side of the Moon had not been seen until a Soviet spacecraft flew around it. That side turned out to be surprisingly different from the one that faces the Earth. The far side has almost no flat plains. It is full of huge craters.

The Moon is about one-fourth the size of Earth. Because it is so small, its gravity is too weak to hold an atmosphere.

The force of gravity affects everything in our Solar System. The gravity of the Sun keeps the planets in orbit. The Earth's gravity keeps the Moon in orbit around it. The Moon's gravity affects the Earth as well. It pulls on our oceans, creating the tides. The tides raise and lower the level of the ocean every 12 hours. We call this high tide and low tide.

Because gravity is the force that holds objects onto the surface of a body in space, if a planet or moon has less gravity than another, a person would weigh less on that planet or moon. Our Moon's gravity is one-sixth what it is on Earth. So if you weighed 100 pounds on Earth, you would only weigh about 16 pounds on the Moon! This weaker gravity allowed the astronauts to walk much more easily on the Moon.

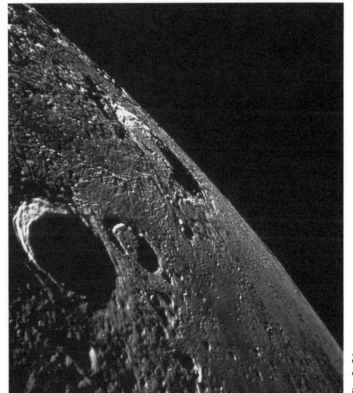

Like the planets, the Moon gives off no light of its own. The light we see radiating from the Moon is actually reflected sunlight. Because the Moon is constantly moving, the direction of sunlight striking its surface changes. This is why it appears full and round at times, and tiny, like a fingernail, at other times.

Scientists are discussing the possibility of building a base on the Moon. This base could be used for scientific experiments and observation of the other planets.

A view of the lunar horizon.
The crater Copernicus is in the background.
The smaller craters in the foreground
are Reinhold and Reinhold-B.

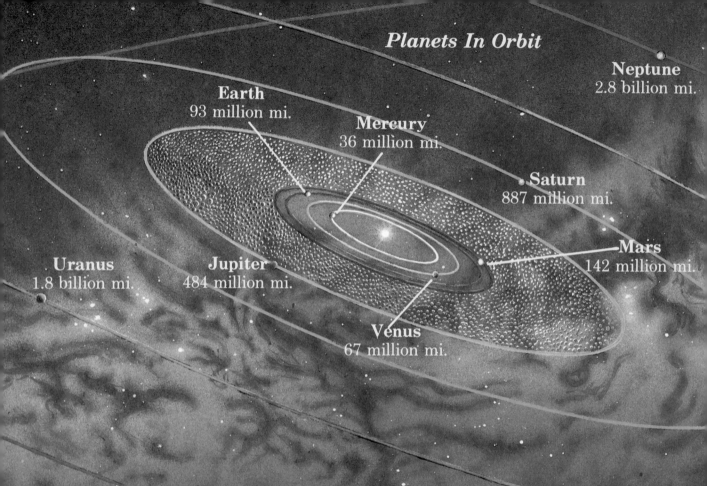

Neptune
2.8 billion mi.

Earth
93 million mi.

Mercury
36 million mi.

Saturn
887 million mi.

Mars
142 million mi.

Uranus
1.8 billion mi.

Jupiter
484 million mi.

Venus
67 million mi.

The Sun is the center of our Solar System. The nine planets and their moons revolve around the Sun in closed paths called orbits. The gravity of the Sun holds the planets in these orbits. Because the Sun is the largest body in our Solar System, it has the strongest gravity. Therefore, it controls the movement of the nine planets.

Closest to the Sun are the four rocky planets, known as the inner planets. These are Mercury, Venus, Earth, and Mars. Beyond Mars is a ring of millions of chunks of rock called asteroids. This ring is known as the asteroid belt. Beyond the asteroid belt are the five outer planets.

These include the four larger planets, which are made up mostly of liquified gases, not solid rock. These four—Jupiter, Saturn, Uranus, and Neptune—are also known as the gas giants. Pluto, the last of the five outer planets, is the farthest from the Sun. It is also the smallest of all nine planets. Pluto has an icy surface, more like the moons of the large outer planets.

An easy way to remember the order of the planets, starting with the one closest to the Sun, is by memorizing the sentence: "My Very Excellent Mother Just Served Us Nine Pies." The first letter of each word in this sentence is also the first letter of one of the planets: Mercury, Venus, Earth, Mars, Jupiter, Saturn, Uranus, Neptune, Pluto.

Pluto
3.7 billion mi.

The distance of each planet from the Sun
is listed below the name, in miles.

Mars

For many years people believed that there might be life on Mars. There is some water on Mars. Most of it exists as frozen ice. At night the temperature on Mars can drop as low as 193°F below zero. Summer temperatures, however, can reach 72°F. Mars also has clouds and fog, volcanoes, lava fields, canyons, and cracks in its crust, just like Earth. Its north and south poles have ice caps, and Mars even appears to have seasons, just like Earth. These conditions led scientists to suspect that there might be life there. In 1976 the Viking space probe landed on Mars and began to send back information about the soil, air, and water. Scientists then concluded that there is no life on Mars.

As similar as Mars and Earth may be, there are also great differences between the two planets. Mars has craters, like the moon. Its atmosphere is thinner than Earth's, and is composed mostly of carbon dioxide. The dust and soil on Mars is red because the iron in it has rusted. This dust is whipped up by enormous dust storms that can sometimes be seen by telescope from Earth. Because of this, Mars is referred to as "the Red Planet."

The Viking 2 spacecraft on a boulder-strewn field of red rocks on Mars' Utopian Plain.

The days and nights on Mars last about as long as they do on Earth. The Martian year, however, lasts for 687 Earth days.

Mars has huge mountains and deep canyons. The extinct Martian volcano called Olympus Mons (Mount Olympus) is the tallest mountain in the Solar System. It is 16 miles high, three times as tall as Mount Everest, the tallest mountain on Earth.

Mars has two small moons. Phobos revolves faster and is the larger of the two. It is 12 miles in diameter and revolves around Mars in only 7½ hours because it is only 3,500 miles away from the planet. Earth's moon is 239,000 miles away from Earth.

Measuring only 10 miles in diameter, Deimos is the smaller of the two Martian moons. It is farther away from the planet and takes 30 hours to complete its journey around Mars.

Mars, once written about in science fiction stories as the home of "little green men" ready to attack the Earth, has the best chance of being the first planet to actually be visited by people from Earth.

Jupiter

Jupiter, Saturn, Uranus, and Neptune are the four largest planets in our Solar System. Because they are made up of liquified gases and do not have solid surfaces like the inner planets, these four outer planets are known as the "gas giants."

Jupiter, the largest planet in our Solar System, was named in ancient times for the supreme god of the Romans. It is so big that 1,000 Earths could fit inside it. Like the Sun, Jupiter is composed mostly of hydrogen. The tremendous gravity of Jupiter has compressed the hydrogen so much that the gas has become a liquid. Because of Jupiter's greater gravity, if you weighed 100 pounds on Earth, you would weigh 254 pounds on Jupiter!

Dark bands appear on Jupiter's surface. These bands are regions where gases like carbon, nitrogen, and phosphorus are sinking. Jupiter's outer atmosphere is very cold, reaching temperatures of 234°F below zero. The temperature at the gaseous center of the planet is a scorching 54,000°F—five times hotter than the surface of the Sun!

Before the Voyager 1 space probe took photos of Jupiter in 1979, astronomers did not know that Jupiter has rings around it. These rings, made of dust, are not visible from Earth.

Jupiter spins very fast. Despite its enormous size, it takes only 9 hours, 50 minutes to complete one rotation. It does, however, take almost 12 years for Jupiter to revolve around the Sun.

Swirling storm clouds and flashing lightning are constantly present on Jupiter. One giant hurricane, known as the Great Red Spot, has been raging for over 300 years. This huge storm system is three times larger than the Earth!

As far as we know, Jupiter has 16 moons. The four largest are known as the Galilean moons—named for the famous astronomer, Galileo, who discovered them in 1610. These are Io, Europa, Callisto, and the largest moon in our entire Solar System, Ganymede.

Ganymede is larger than the planet Mercury. Its surface is covered with cracked ice and rocks. Europa and Callisto are also covered with ice. Callisto is the Galilean moon that orbits farthest from Jupiter, more than one million miles from the planet. Io looks different from the other Galilean moons. It has the Solar System's only active volcanoes, other than the ones on Earth. The volcanoes send hot material from deep inside the moon out onto its surface, giving Io its bright colors.

The Great Red Spot
as seen from a distance of
1,636,148 miles
by the Voyager 2 spacecraft.

Ganymede,
the largest
moon in our
Solar System.

17

Saturn

For many years Saturn was called "the ringed planet." Its rings are relatively easy to see from Earth with a small telescope. When Galileo discovered the rings in 1610, he didn't know what they were. They first were identified as rings by the astronomer Huygens in 1655.

The rings of Saturn are made up of ice particles. Some of these particles are just specks; others are large boulders. There are seven main rings made up of thousands of smaller ringlets that revolve around Saturn.

Like the other gas giants, Saturn is mostly made up of liquified hydrogen and helium gas, mixed with ice. It is nearly 800 times bigger than the Earth. It takes Saturn only 10 hours, 14 minutes to complete one rotation, but it takes almost 30 years for the planet to complete an orbit of the Sun.

Saturn has at least 17 moons, the most of any planet in our Solar System. Many of these moons are simply "giant snowballs" less than 60 miles in diameter. Saturn's larger moons are: Mimas, Enceladus, Tethys, Dione, Rhea, Titan, Hyperion, Iapetus, and Phoebe.

Titan is the largest of Saturn's moons. Its atmosphere, ten times as thick as Earth's, is made up of nitrogen and methane. The Voyager space probe could not "see" through this thick, foggy atmosphere. Like the clouds on Venus, this gaseous blanket holds in the heat, making Titan a relatively warm moon.

The surface of another moon, Mimas, is covered with craters. One huge crater, named Herschel, is about 80 miles wide. Hyperion, Iapetus, and Phoebe, the outer three moons of Saturn, were probably not originally moons. They are believed to have originally existed as asteroids or comets that came within Saturn's gravitational pull and began to orbit the planet. Phoebe, the outermost moon, orbits in the opposite direction of the other of Saturn's moons.

A montage of photographic images showing Saturn with some of its moons. Dione is the moon in front of Saturn.

Uranus

Uranus was discovered by an amateur astronomer, named Sir William Herschel, in 1781. In 1986 Voyager 2 passed by Uranus, giving us photos and new information about this planet.

In 1977 astronomers discovered that Uranus has rings. The Voyager photos gave us our first close-up view of Uranus' 15 rings.

Uranus is covered by a blue-tinted, frozen fog that has a temperature of 360°F below zero. It takes only 17¼ hours for Uranus to rotate, but it takes 84 years for it to go around the Sun.

The most unusual feature of Uranus is that it spins on its side, like a top that has fallen over. Because of this, one of the poles (north or south) is always facing the Sun. During its 84-year trip around the Sun, each pole remains in sunlight for 42 years.

Uranus has 15 moons that we know of. Most of the moons are charcoal-black chunks of rock and ice. They are quite small, averaging about 40 miles in diameter. The five largest moons are: Miranda, Ariel, Umbriel, Titania, and Oberon.

An artist's concept of Uranus and its rings.

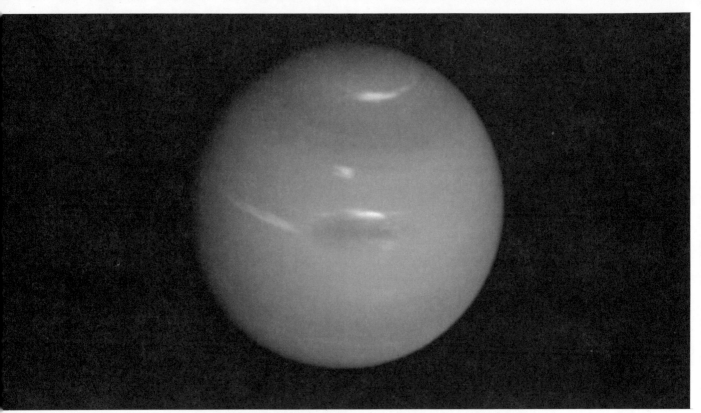

The Great Dark Spot (GDS), seen at the center, is about the size of Earth.
The "Scooter" is the small, bright cloud below the GDS.

Neptune

The last of the gas giants is Neptune. This frozen, gaseous planet was discovered in 1846 by the German astronomer, Johann Gottfried Galle.

Neptune is four times larger than Earth and 2.7 billion miles away from it. It takes 165 years for Neptune to orbit the sun, and about 16 hours to complete one rotation.

In 1989, the Voyager 2 spacecraft passed about 3,000 miles above Neptune's north pole. Voyager discovered several faint rings encircling the planet, and six new moons.

Despite the cold of −328F, neptune has a turbulent weather system. Voyager measured 1,300 mph winds blowing wildly over the blue-green planet. A huge storm, called the Great Dark Spot (GDS), can be seen on the planet's surface, and a quick-changing cloud feature has been dubbed the "Scooter."

Triton, one of Neptune's two previously known moons, was Voyager's last encounter before heading for outer space. It may have been a planet that was captured by Neptune's gravity. Triton is one of only two moons in our solar system that has an atmosphere. When the ice within Triton is warmed, it erupts into five-mile high geysers of black snow.

Pluto

Pluto is the smallest, coldest, and farthest planet from the Sun. It takes 6 days, 9 hours for Pluto to rotate, and 248 years for it to complete a revolution around the Sun. Pluto is so cold that scientists believe that it is a giant ball of ice. Pluto's temperature is about 400°F below zero.

Pluto was discovered in 1930 by an American astronomer named Clyde Tombaugh. Some astronomers believe that Pluto was once a moon of Neptune, which was pulled out of orbit by the gravity of still another —as yet undiscovered—planet. Others believe that Pluto fell into orbit in our Solar System while traveling through space, and was caught in the Sun's gravitational pull.

For a time some scientists believed that Pluto and its moon, Charon, might have actually been twin planets, rather than a planet and a moon. We now know that Charon is definitely a moon. Charon is about one-half the size of Pluto. They travel around each other in just under a week at a distance that is 20 times closer than the distance between the Earth and its moon.

Because Pluto is so small, it has very little gravity. If you weighed 100 pounds on Earth, you would weigh only 5 pounds on Pluto!

An artist's view of Pluto, its moon Charon, and in the distance, our Sun.

Asteroids

Between the orbits of Mars and Jupiter is a "belt" of thousands of dwarf or minor planets called asteroids. Asteroids are giant boulders, small chunks of rock and pieces of metal, all of which move through space. Like planets, they revolve around the Sun and reflect light. The largest asteroid is Ceres, over 600 miles wide. Not all asteroids are in this asteroid belt. Some—namely, Eros, Apollo, and Amor—pass very near to Earth. Asteroids may be the remaining chunks of a planet that fell apart long ago. Or they may be the pieces of a planet that was about to form, but never did.

Comets

Far beyond Pluto is a body called the Oort Cloud. This is a huge cloud of ice and dust, surrounding our Solar System. Every once in a while, a passing star shakes up the Oort Cloud, sending chunks of it hurling toward our Sun. These chunks become comets. A comet is a big ball of frozen gas and rock, often described as a "dirty snowball in space." As the comet gets closer to the Sun, the frozen gas melts. The stream that this melted gas creates trails behind the comet, forming its tail. When a comet passes close enough to Earth, we can see the tail spreading across the night sky. The most famous comet is Halley's Comet. Astronomer Edmund Halley proved that comets seen in the years 1531, 1607, and 1682 were all the same comet. He then predicted that the same comet would return in 1758. When it did (16 years after his death), the comet was given his name. Halley's comet will next be seen here in the year 2062.

Meteors

Sometimes as comets pass the Sun, small pieces break off. When these pieces of comets enter the Earth's atmosphere, they burn up because of the heat created by friction with the air. These burning rocks are called meteors. The "shooting stars" we sometimes see in the night sky are not stars at all. They are meteors burning up in the Earth's atmosphere. In addition to pieces of comets, meteors can also be small asteroids from the asteroid belt whose orbits intersect with Earth's orbit. Sometimes a meteor does not burn up completely when it hits the Earth's atmosphere. These land on Earth and are studied by scientists. Meteors that land on Earth are called meteorites.

The Milky Way

Our Solar System is a group of planets which revolve around one star, our Sun. Our Sun is one star in a collection of hundreds of billions of stars. This collection of stars is called a galaxy. The name of our galaxy is the Milky Way. The Milky Way is shaped like a giant whirlpool. To understand the shape of a whirlpool, watch the water drain out of your sink or bathtub when you pull the plug. The shape made by the water as it goes down the drain is a whirlpool. Another name for a whirlpool galaxy is a spiral galaxy. To understand the shape of a spiral, take a piece of paper and a pencil. Start on the edge of the paper and draw a long, continuous circle that goes around and around, moving toward the middle of the paper.

The Milky Way galaxy is 100,000 light-years in diameter. A light-year is a measure of distance. One light-year is the distance you would cover if you traveled at the speed of light (186,000 miles per second) for one year.

Our Sun is located two-thirds of the way to the edge of the Milky Way galaxy. All of the stars we see in the night sky on Earth are part of our Milky Way galaxy. There are billions of galaxies in the universe. Galaxies group together into clusters. The Milky Way is part of what we call the Local Group, which contains about 24 galaxies. The Milky Way is the second largest of these galaxies. The largest of all is the Andromeda Galaxy, a spiral galaxy which is 2 million light-years away from Earth, and can just barely be seen without a telescope.